Fluid Power
Educational
Series

I0491821

Electro-hydraulic Servo Valves

Joji Parambath

Electro-hydraulic Servo Valves

Copyright © 2022 Joji Parambath

All rights reserved

ISBN: 9798654202642

https://jojibooks.com

First Edition: 2020
Revised Edition: 2022

Disclaimer of Liability

The contents of this book have been checked for accuracy. Since deviations cannot be precluded entirely, we cannot guarantee full agreement. Only qualified personnel should be allowed to install and work hydraulic equipment. Qualified persons are defined as persons who are authorized to commission, ground, and tag circuits, equipment, and systems following established safety practices and standards.

Dedicated to

all my former colleagues at Foremen Training Institute,
Bangalore

Table of Contents

PREFACE

Servo valves were developed in the late 1930s as a high-tech, high-performance, but high-cost solution to the motion control requirements. A servo valve uses a torque motor to place its spool at the desired position. It is used mainly for the closed-loop control of position or speed or force. High-performance closed-loop servo valve technology has become the norm in machine automation, where the requirements are greater precision, faster operation, and more straightforward adjustments.

This book explores the technology used in state-of-the-art servo valves. The book also describes the construction of electro-hydraulic servo valve systems, the details of various types of control elements, and the characteristics of servo valve systems.

Many other fluid power topics are given in other textbooks under the fluid power educational series by the same author. A list of all the textbooks is given at the end of the book (Page Nos. 54-55). Also, please see the details at https://jojibooks.com

Enjoy reading the book.

Your feedback is most welcome.

JOJI Parambath

Chapter 1 | Introduction to Electro-hydraulic Valves

When a discrete type electro-hydraulic directional control valve operated by the conventional on/off solenoids is actuated, its spool is pushed entirely over to its maximum travel, usually against the restraining force of a spring. In other words, its spool can only be set in two or three discrete positions.

The precise control of the position, speed, pressure or force is difficult to achieve in systems using the discrete electro-hydraulic valves, due to the 'jerkiness' of motion that these valves experience. Moreover, the remote control of pressure or flow rate is not possible when using these valves. Therefore, the conventional type solenoid valves are primarily used in less sophisticated systems to get the direction control of the flow. More and more electronics are integrated into the traditional hydraulic valves to improve their accuracy and performance.

Electro-hydraulic valves operated by electronic controllers have been developed to overcome the problems that the discrete solenoid valves encounters and to obtain the automated step-less control of pressure and/or flow rate.

Infinitely Variable Hydraulic Valves
In an infinitely variable electro-hydraulic valve, the valve output can be infinitely varied with the input signal applied to the valve. The valve's response needs to be proportional to the input command signal applied to it. The input signal may correspond to the required position, velocity, or force required in the system. That is, the valve can provide closed-loop control with exact positional accuracy, repeatable velocity profiles, and predictable force or torque regulation.

The servo valves are classified into the following two types: (1) Proportional valves and (2) Servo valves.

.

Discrete Valves Vs Infinitely Variable Valves

The most common valve configurations used for hydraulic control systems range from the discrete valves controlled by the electromagnets to infinitely variable valves, controlled by the special proportional electromagnets or torque motors. A comparison of these valves is most appropriate for broadening the concepts and is given in Table 1.1.

Table 1.1 | Discrete valves Vs infinitely variable valves

Discrete valves	Infinitely variable valves (Proportional & Servo valves)
• Mainly used in open-loop control systems	• Mainly used in closed-loop control systems
• Use conventional on/off solenoids	• Use proportional solenoids or torque motors
• The spool moves to its full stroke	• The spool moves in relation to the input signal
• Do not require spool position sensing	• Require spool position (output) sensing
• Provide slow response	• Provide fast response
• Less accurate	• More accurate
• Cheaper	• Expensive

Proportional Valves Vs Servo Valves

The distinction between the proportional valves and the servo valves is inconsistently stated, but, in general, the servo valves provide a higher degree of closed-loop control.

A proportional valve can be seen as a conventional directional control electro-hydraulic valve tailored for obtaining proportional characteristics and is gradually acquiring more and more servo valve characteristics.

Though the objective of the servo valves and the proportional valves are similar, the servo valves are more accurate than the proportional valves.

Chapter 2 | Mechanical Type Servo valve

Figure 2.1 shows a typical closed-loop mechanical servo valve system with a cylinder and a normally-closed control valve. The valve has a sliding sleeve that controls the flow to the cylinder with the help of a feedback link attached to the cylinder.

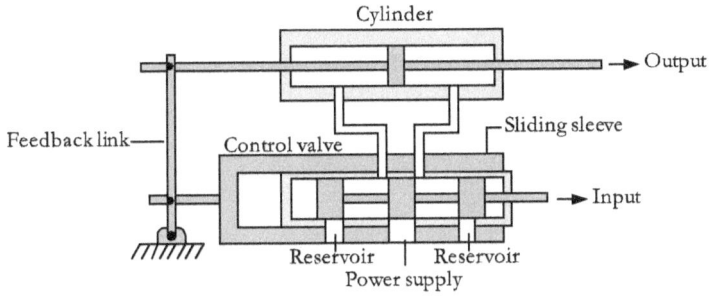

Figure 2.1 | A mechanical servo system

The input signal in the form of a small force applied to the valve shifts the spool by a specified length of travel and opens the valve. Therefore, the fluid flows into the cylinder through the valve and produces an output motion of the cylinder. The movement of the cylinder piston causes the feedback link to move the sliding sleeve until it blocks the flow to the cylinder again. It can be seen that a given amount of displacement of the spool produces a corresponding amount of the output motion of the cylinder.

A common category of applications of mechanical type servo valves is in the hydraulic power steering systems of automobiles and other transport vehicles. However, the mechanical type servo systems were found to be unsuitable for high-performance complex systems. Therefore, electro-hydraulic servo systems with electronic controllers have been devised to obtain increased accuracy and additional flexibility.

Chapter 3 | Electro-hydraulic Servo Valve System

A servo valve system is a high-performance closed-loop system used for the precise control of the output (load) parameter, such as its position, velocity, or force, in response to a command signal. The block diagram of Figure 3.1 shows the electro-hydraulic servo valve system. It primarily consists of a transducer, a servo amplifier, a servo valve with a torque motor and an actuator with a connected load.

The output of the servo system is measured and converted into an equivalent electrical signal by the transducer. This signal acts as the feedback signal and is compared to the command signal (representing the required output) to generate an error signal. The error signal is modulated and amplified by the amplifier. The amplified signal is used to control the servo valve. The servo valve regulates the fluid flow to the actuator in proportion to the current from the amplifier. The actuator then drives the load to move in such a precise manner as to reduce the error signal.

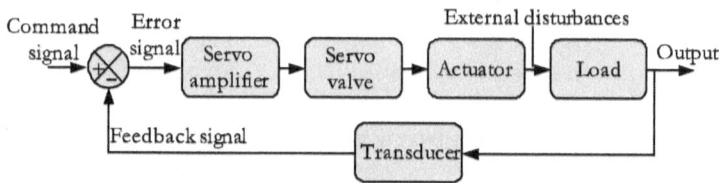

Figure 3.1 | A block diagram of a servo valve system

According to the type of feedback signals, the servo systems can be classified into three basic types. They are: (1) Position servo systems, (2) Velocity servo systems, and (3) Pressure/Force servo systems. The servo valves can be interfaced with PLCs or HMIs. However, they are expensive as compared to other types of hydraulic valves. Moreover, they are prone to contamination due to their minimal tolerances.

Chapter 4 | Construction of Servo Valves

The high precision offered by servo valves is achievable with the right selection of materials, manufacturing techniques, and mechanical designs. A hydraulic servo valve system is constructed with a servo valve with a torque motor and a spool shifting mechanism and a servo amplifier.

The mechanical design of the servo valves, their materials of construction, and their general workmanship should be compatible with the intended operational, environmental, and service life requirements for the servo valves.

Another critical consideration in servo valves is the lap condition at the valve centre (null region). Select spool cuts are to be used if the performance in the null region is critical.

Servo Amplifier

As stated earlier, the servo amplifier amplifies the error signal in the electro-hydraulic servo valve system. Figure 4.1 shows the symbolic representation of the servo amplifier. It mainly contains a summing junction, an amplifier with a proportional, integral, and derivative (PID) circuit, a ramp generator, and a dither oscillator.

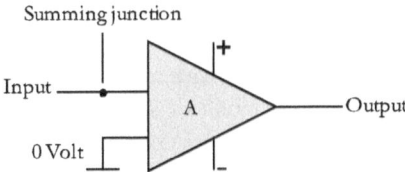

Figure 4.1 | A schematic diagram of a servo amplifier

The servo amplifier generates an error signal, due to the difference between the measured output signal and the command input signal, at the summing junction. It has very high input impedance and, therefore, minimal current flows into the amplifier input. The amplifier is provided with an integrated PID control circuit for the optimum performance of the closed-loop control circuit. See Appendix 2 for the concepts of PID controls. The servo amplifier provides a control current typically of the order of milliamps to drive the servo valve.

The servo valve responds to the amplifier signal by controlling the fluid delivery into/out of the associated actuator. The ramp generator is used to control the rate of change of the amplifier's output and consequently controls the rate at which the valve opens or closes. The dither oscillator is used in the amplifier to reduce the effects of friction due to the sliding of the spool against the valve body, and inertia.

Proportional amplifiers must accommodate LVDTs that require signal conditioners. Therefore, servo amplifiers tend to be much simpler than proportional valve amplifiers.

Typically, the torque motors operate with less than 100 mA, and they can be shifted even with a few milliwatts (mW) of power.

Servo Valve
The servo valve is the critical component of a servo valve system. It is a precisely machined spool-type directional control valve with a current-driven mechanism using a torque motor to control the spool position of the valve.

The torque motor can be considered as an electromechanical transducer that produces a small deflection in proportion to the input current. It can impart the necessary motion to the spool either directly or indirectly.

The spool can be infinitely positioned to control the direction of the fluid flow and the amount of pressure in a hydraulic system, in response to the electrical/ electronic control signals.

The fluid to the first stage is filtered through an internal filter.

Designs for servo valves vary widely to meet varied application requirements. A fundamental distinction in servo valves is the number of stages of hydraulic amplification involved. Accordingly, the servo valves are available in one-stage, two-stage, or three-stage, designs.

Single-stage Servo Valve
The single-stage servo valve consists of a torque motor directly attached to the output stage of the valve that is usually a sliding spool. The torque motor must supply enough torque to the valve to shift its spool directly against an opposing pressure.

The single-stage valve has limited power capability, and it is also susceptible to stability problems. Figure 4.2 shows the block diagram of the single-stage servo.

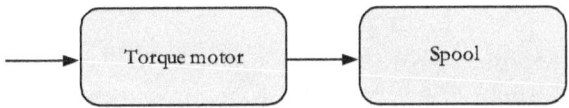

Figure 4.2 | A single-stage electro-hydraulic servo valve

Two-stage Servo Valve
Two-stage servo valves are the most common type of servo valves used in electro-hydraulic systems. A two-stage servo valve uses a pilot spool to enhance its hydraulic amplification.

The two-stage servo valve consists of a torque motor, a pilot spool, and the main spool, as shown in Figure 4.3. First, the torque motor shifts the pilot spool, which in turn, directs fluid flow to move the main spool.

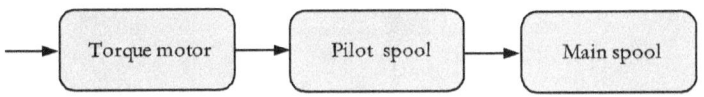

Figure 4.3 | A two-stage electro-hydraulic servo valve

Three-stage Servo Valve

The three-stage servo valve is similar to the two-stage servo valve, except that the pilot stage itself is further divided into two stages, namely 1st stage pilot, and 2nd stage pilot, as shown in Figure 4.4. The first-stage pilot spool shifts the second-stage pilot spool, which, in turn, shifts the main spool.

The three-stage servo valves are used for applications with high-flow, high-pressure requirements. The steel industry process applications remain the primary domain of the three-stage high-performance servo valves.

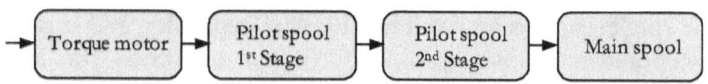

Figure 4.4 | A three-stage electro-hydraulic servo valve

Torque Motor

The main spool in a servo valve is required to be shifted to its correct position to get the desired change in the output of the valve. This change can be achieved by using an electromechanical torque motor at the input stage of the servo valve. It produces a small deflection in proportion to the input current and imparts the necessary motion to the spool, either directly or indirectly.

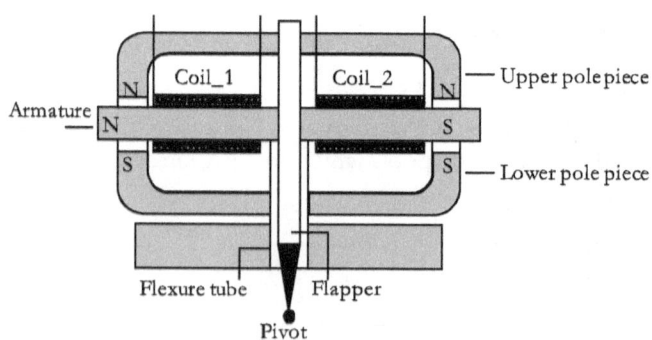

Figure 4.5 | A torque motor of a servo valve

The torque motor consists of upper and lower pole pieces, an armature, two coils, a flapper and a flexure tube, as shown in Figure 4.5. The two pole-pieces – one polarised 'North' and the other polarised 'South' - provide paths for the magnetic flux.

The armature is mounted on a pivot and is suspended in the air gaps between the upper and the lower pole pieces. The ends of the armature are positioned in the middle of the air gap.

When electric currents flow through these coils, the armature gets polarised. Therefore, one end of the armature is attracted to one pole piece, and the other end is repelled by the same pole piece at the other end.

In this way, the torque motor converts the input signal from the servo valve's controller to a proportionate semi-rotary movement of the armature. The displacement of the armature is limited to a tiny value, typically, a few thousandths of an inch.

When the armature polarisation is reversed by reversing the polarity of the input current, the rotational torque of the armature is also reversed.

The small deflection of the armature can be used to control the pilot stage of the servo valve through the connecting mechanism, such as the flapper.

The armature/flapper assembly is supported on the flexure tube that allows it to rotate until the torque produced equals the restraining torque in the flexure tube. The arrangement of the armature/flapper assembly imparts a controlled motion to the spool of the servo valve and positions the spool in proportion to the input current.

It may be noted that the torque motor produces a much more linear current/force relationship than that provided by the proportional solenoid.

Spool Shifting Methods in servo valves

The torque motor in a servo valve can shift its spool directly or indirectly for the hydraulic power amplification. Although the power of the flapper lever is minuscule, the hydraulic power amplification is enough to position the spool accurately.

Two primary forms of construction are used for connecting the armature of the torque motor to the pilot stage of the valve. They are: (1) Flapper nozzle arrangement and (2) Jet pipe arrangement.

Flapper Nozzle Arrangement

Servo valves with the flapper nozzle arrangement are the most commonly used type of servo valves in the industry today. The flapper nozzle arrangement consists of a flapper lever and two nozzles, as shown in Figure 4.6. This arrangement can be considered as the first stage of a two-stage servo valve.

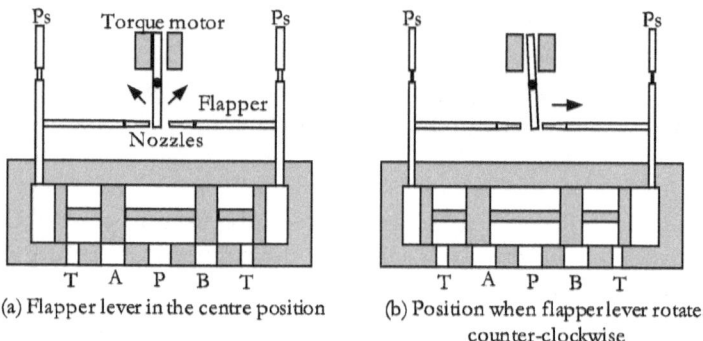

(a) Flapper lever in the centre position   (b) Position when flapper lever rotates counter-clockwise

Figure 4.6 | Two positions of the servo valve with a flapper nozzle arrangement

The flapper lever, attached to the armature of the torque motor, is aligned between the two nozzles, as shown in Figure 4.6(a). This arrangement forms two variable orifices between the nozzle tips and the flapper. These nozzles are very critically sized so that the valve works as designed. The currents in the armature coils produce a magnetic field that deflects the flapper lever clockwise or anticlockwise, depending on the magnetic polarity. The movement of the flapper lever controls the flow through the nozzles.

If the torque motor rotates the flapper counter-clockwise, as illustrated in Figure 4.6(b), then the orifice on the right-hand side becomes smaller than that on the left-hand side. The resulting pressure on the right side of the spool becomes higher than that on its left side. This pressure difference moves the spool to the left by an amount proportional to the current levels in the armature coils.

The position of the spool controls the amount, rate, and duration of the flow to the actuator.

As the spool moves in response to the pressure differential across the spool, a feedback wire exerts an opposing torque on the armature/flapper assembly trying to re-centre the flapper. The spool continues to move until the torque produced by the feedback wire equals that produced by the control signal.

Jet Pipe Arrangement

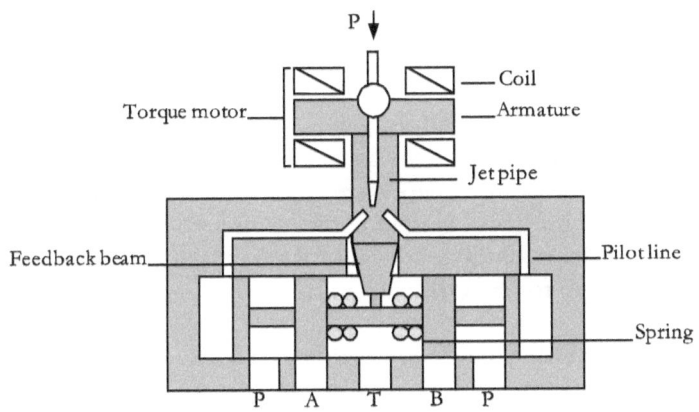

Figure 4.7 | A servo valve with a Jet pipe

A jet pipe arrangement in a servo valve system consists of a jet pipe with a specially shaped nozzle, two pilot lines, and a spring-biased main spool, as shown in Figure 4.7.

The jet pipe with a nozzle is attached to the armature of the torque motor. It is also mechanically linked to the main spool of the valve. When the nozzle is applied with a pilot pressure, equal amounts of fluid are directed to the pilot line receivers.

When there is no error signal from the servo valve system, the nozzle is centred between the pilot lines.

However, the presence of an error signal displaces the jet pipe from its neutral position. This displacement of the nozzle directs more fluid into one pilot line than into the other line. This imbalance of the flow causes the development of unequal pressures at the ends of the main spool. As a consequence, the spool shifts to reduce the error signal.

At the same time, the jet pipe, which is mechanically linked to the spool, tends to align centrally between the two pilot lines.

The servo valve design with the jet pipe is less sensitive to contamination than the design with the flapper nozzle.

Lap Conditions in Servo Valves
The lap condition of a servo valve is an essential parameter for the valve. The type of the valve centre in the servo valve is defined by the width of the spool land in comparison with the width of the port in the sleeve of the servo valve. It affects the amount of spool travel required to open the valve from its centre position.

Three critical lap conditions exist in servo valves. These are (1) Line-to-line (zero-overlap), (2) Open-centre (under-lapped), and (3) Closed-centre (over-lapped). Figure 4.8(a), (b), and (c) respectively illustrate these lap conditions.

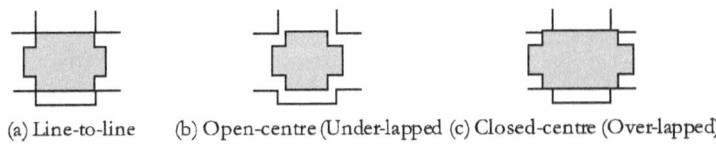

(a) Line-to-line   (b) Open-centre (Under-lapped   (c) Closed-centre (Over-lapped)

Figure 4.8 | The lap conditions in servo valves

Line-to-Line Configuration

The most commonly used type of lap condition in servo valves is the line-to-line configuration [Figure 4.8(a)], where the spool of each valve is machined and aligned very precisely to obtain the line-to-line fit of the flow metering edges with the opening in the valve body.

Therefore, the zero-lapped valve is capable of delivering the flow to the connected actuator ports immediately, in response to a differentially small amount of spool shift, either way. This type of construction provides a higher pressure gain.

Under-lapped Configuration

In a servo valve with the under-lapped (open-centre) design [Figure 4.8(b)], the land width of the spool is smaller than the port width of the valve. Therefore, the under-lapped valve delivers the flow even when the valve remains in its centre position. The lap condition for the under-lapped valve provides for a lower pressure gain.

Over-lapped Configuration

In a servo valve with the over-lapped (closed-centre) design [Figure 4.8(c)], the land width of the spool is greater than the port width of the valve. The lap condition for the overlapped servo valve provides for the same pressure gain as that of the line-to-line lap condition. However, the over-lapped valve has a 'deadband'. It means that the spool has to move a certain distance before the flow can be delivered. The maximum amount of overlap for a servo valve is approximately 2%.

Effect of Contamination on Servo Valves

The servo valve clearances are tiny, and, therefore, they are extremely prone to fluid contamination. The contamination can gum up the valve and results in its reduced response.

It also increases the amount of wear of the spool which will, in turn, reduce the accuracy of the valve.

Further, it can also cause the small opening of the nozzle to be reduced, and the speed of the fluid across the nozzle can reach high velocities.

It also causes the wear of the nozzle, and as a consequence, it affects the null condition of the valve. The net result is an inferior performance of the servo valve system.

Null Drift and Null Adjustment

In the null position of the servo valve, the spool of the valve remains at the reference centre position. There is a tendency for a servo valve especially the zero-lapped one, connected to a hydraulic system, to drift from its centre position due to the presence of contamination, or due to the variations in temperature or supply pressure or load pressure in the system.

The shift in the null position of a servo valve is indicated when the associated actuator experiences significant flow at zero input signal and causes errors in the system.

Therefore, it is often necessary to readjust the null position of the servo valve to make the flow output of the valve zero or make the pressures at the blocked working ports of the valve equal, when no input signal is applied. This adjustment can be carried out by turning a mechanical null screw provided on the valve. It is an eccentric bushing retainer pin that when rotated provides the null adjustment. The servo valves typically allow $\pm 20\%$ adjustment of the flow null.

Chapter 5 | Control Elements of Servo Valves

Potentiometer

A potentiometer is used to set the required spool position in a servo valve. Figure 5.1 shows the two alternative schematic representations of the potentiometer. It consists of a resistor and a wiper placed on the resistor. The position of the wiper determines the voltage (V_o) at the wiper terminal. The voltage must be linearly related to the position of the wiper. The ideal function of the potentiometer is to provide a command input voltage to the connected amplifier.

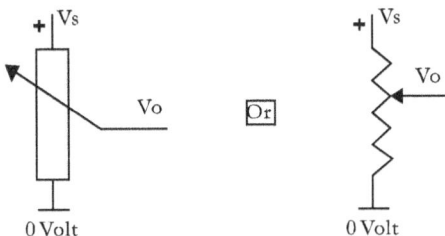

Figure 5.1 | Schematic diagrams of the potentiometer.

However, in practice, a signal voltage at higher level results in the development of a higher voltage drop across the upper part of the potentiometer, and a consequent reduction in the output voltage of the potentiometer. This voltage drop gives rise to a non-linear relationship between the wiper position and the wiper terminal voltage V_o.

The non-linear relationship in a potentiometer can be reduced by selecting an appropriate potentiometer resistance. As a rule of thumb, the maximum resistance of the potentiometer should be about 10% of the input resistance of the associated amplifier, to restrict the non-linearity of the potentiometer within 2 to 3%.

Joystick

A joystick is a mechanism to translate the movement of a stick into electronic information for the control of machines, such as cranes and robots. The rotational positions of the joystick can be measured using potentiometers, optical or magnetic sensors, and incremental encoders. The measured values correspond to the positions of the joystick and act as the command input signals to the connected amplifier. The most critical parameters of the joystick are its linearity, resolution, size, and cost.

Electronic Control Unit

The electronic control unit in a servo valve system is the brain of the system. Figure 5.2 shows the functional schematic of a typical control unit.

An essential element in the control unit is an amplifier that converts the low-power input signal into an output signal sufficient to operate a servo valve. The input signal may come from several devices, including a potentiometer or a joystick.

The control unit for a closed-loop servo valve system is designed to receive the spool position feedback from a position sensor.

The electronic control units also include many features, such as ramping (acceleration and deceleration limiting), dither, pulse width modulation (PWM), and current sensing circuitry to control the associated servo valves accurately.

The output current in the control unit is usually modified using the ramping, dither, and PWM techniques. The maximum output current is limited by setting the potentiometer for the 'Current maximum, I_{max}' to prevent the overdriving of the valve, reduce the effects of temperature change on the solenoid, and protect the valve against electrical short circuits.

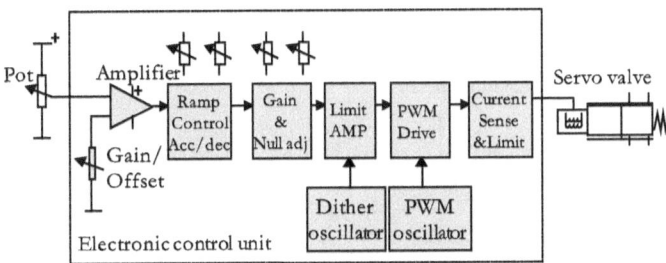

Figure 5.2 | A block diagram of an electronic control unit

The command signals in servo valves may be bipolar voltages (Ex: $\pm10V$) or unipolar voltages (Ex: 0-10 V) or currents (Ex: 4 -20 mA). In many applications, the current input signal is preferred to the voltage input signal to avoid the effects of large voltage drops across long control lines. Typically, 4 mA represents the zero signal, and 20 mA represents the maximum signal. Any current below 4 mA is ignored to make the system less sensitive to the electrical noise or to provide an indication of the system malfunction. The following paragraph discusses the fundamental principles of electronic amplifiers.

Servo Amplifier
The servo amplifier is designed to convert the low-power electrical command signal into a high-power current signal to run the associated servo valve. As the input signal for controlling the servo valve comes from a low power device, such as a potentiometer or a sensor or a transducer, it is necessary to amplify the input signal to operate the servo valve. An electronic amplifier using a transistor is capable of producing a large current flow through its output in response to a small voltage at its input. Figure 5.3 shows the schematic diagram of the amplifier with input and output terminals and a summing junction.

An important parameter of the servo amplifier is the open-loop gain (A). It is the ratio of its output voltage to its input voltage. Remember that the amplifier inverts the input signal. Therefore, the output voltage can be given by the formula:

Output voltage = -A x Input voltage

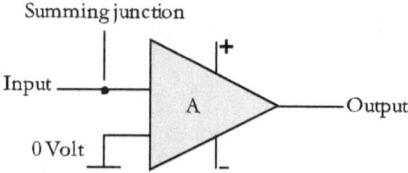

Figure 5.3 | A schematic diagram of a servo amplifier

The following bulleted list the essential characteristics required of the servo amplifier:

- The amplifier's input impedance must be high so that only a small current should enter the amplifier from the input.
- The amplifier's output impedance must be small so that a large current can be drawn from the output, keeping the voltage constant.
- The summing junction is considered to be always at zero potential.

The type of amplifiers used in control systems is typically referred to as operational amplifiers (OP-AMPs). It is a high-gain amplifier whose operational characteristics are determined by the use of external feedback elements, as shown in Figure 5.4.

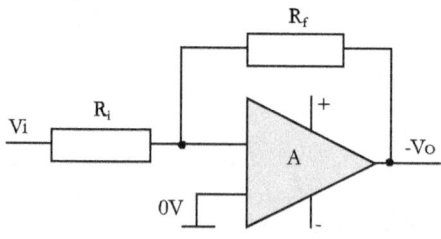

Figure 5.4 | An amplifier with a feedback element

For the amplifier shown in Figure 5.4,

$$V_o = - \frac{R_f}{R_i} x V_i$$

The ratio (R_f/R_i) is the closed-loop gain (A) of the amplifier. As can be seen, the amplifier's gain depends on the value of R_i and R_f.

Additional Features of Electronic Control Unit

In a servo valve, certain additional functions need to be carried out. That is: it may be necessary to control the valve's opening and closing rates. There is also the need to control the analog type solenoid coil of the valve digitally and overcome the problem of spring hysteresis, stiction, and inertia. Therefore, the electronic control unit of the valve uses various techniques, such as ramping, pulse width modulation (PWM), and dithering to achieve these additional functions. The following sections explain these techniques.

Ramp Rate Adjustment

The rate of change of the amplifier's output can be controlled through the ramping technique. This method is used to limit its acceleration or deceleration. The ramp rate adjustment ultimately restricts the rate at which the valve opens or closes.

Pulse Width Modulation (PWM)

A primary method for converting the supply voltage to the corresponding output current is by using a rheostat. However, this approach tends to develop a considerable amount of heat and noise.

Therefore, the efficient method employed in the servo valve for converting the supply voltage to the corresponding output current is by using the pulse width modulation (PWM).

It is a current control method in which a transistor located on the amplifier card of the servo valve turns the current flowing through the solenoid coil on and off very rapidly to achieve the desired current flow.

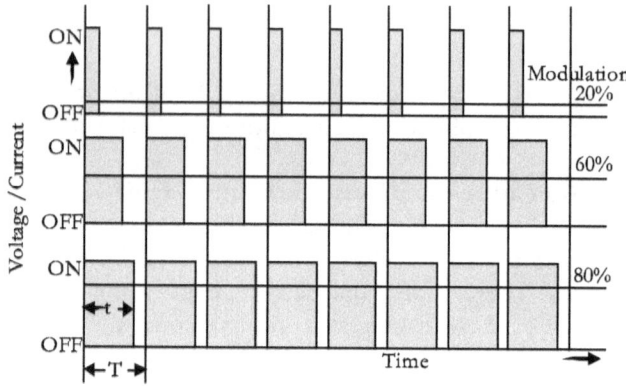

Figure 5.5 | A timing diagram for illustrating the pulse width modulation (PWM) technique

Figure 5.5 shows the timing diagram for illustrating the pulse width modulation (PWM) technique. The displacement of the poppet or spool of the valve can be controlled by varying its 'ON time'. That means if the width of the pulse is 60% of its maximum duration, theoretically the valve shifts enough to deliver a 60% output.

The following texts give essential terms of the PWM technique.

(1) PWM period (T) is the time duration of one PWM cycle.

(2) PWM frequency (1/T) is the rate at which the PWM cycles turn on and off in one second, and is given in Hertz (Hz).

(3) The term pulse width (t) is used to indicate the time during which one PWM cycle remains in the "ON" state during one PWM cycle.

(4) The ratio of the ON time of the PWM to its period (i.e., t/T) is regarded as its duty cycle.

The PWM frequency can be low or high. The low-frequency PWM is in the range from 100 to 400 Hz while the high-frequency PWM is in the range from 4000 to 5000 Hz.

Remember, the PWM frequency must be significantly higher than the frequency that the associated hydro-mechanical valve can respond to. The valve and hence the actuator can respond only to the average current instead of the instantaneous current.

The high-frequency PWM produces a ripple-free amperage output. The PWM has become the power amplification method for controlling the solenoids often requiring up to 4 A of current and as much as 30 Watt of power.

Dither Oscillator
A servo valve operates with small forces derived from the associated solenoid and depends on the small deflections of the valve spool. It is, hence, vulnerable to the problem of spring hysteresis, inertia, and static friction (stiction) between the spool and the body of the valve. The stiction affects the performance of the valve as it causes the valve to ignore small changes in the demanded spool position. The stiction effect is made worse if the valve's spool is held in a fixed position for an extended period,

allowing the spool to settle. The dirt in the fluid medium also encourages stiction.

The effects of the hysteresis, stiction, and inertia can be reduced by superimposing a dither signal to the PWM command signal. The dither signal is a low-amplitude, high-frequency sinusoidal (AC) signal (50 – 100 Hz) used for keeping the valve spool in constant motion around its desired position. This signal is too fast for the spool to follow, but the small spool movements are enough to prevent it from staying in a fixed position.

The dither plays a significant role in improving the hysteresis, response, and stability of the valve. Ideally, the oscillation caused by the dither signal does not alter the output of the servo valve.

The dither in a servo valve is specified by its frequency (Hz) and peak-to-peak current (mA). The dither amplitude is usually adjustable from 0 to 10% of the rated maximum current of the solenoid. The dither amplitude and the frequency are usually factory set.

Chapter 6 | Characteristics of Servo Valves

In the analysis and design of an electro-hydraulic servo valve system, it is essential to design controllers such that a satisfactory response is obtained at all times. The system response has two essential components. They are: (1) Transient (dynamic) response and (2) Steady-state response. The dynamic and steady-state characteristics of the servo valve system are evaluated to study its performance accurately.

The transient response of the servo valve system represents the momentary variations of the system parameters, such as the pressure and the flow rate, in response to a sudden change in the input signal applied to the system. The transient response is usually present for a short duration of time, immediately after the system is turned on. If the system is inherently stable, the transient response will disappear soon. As we see later, the transient response of a servo valve can be described by its 'step response' and 'frequency response'.

The steady-state condition of the servo valve system is a state in which system parameters, such as the pressure and the flow rate, do not vary significantly with time after the initial fluctuations of the parameters have disappeared. For example, the steady-state speed of a hydraulic cylinder can be calculated by dividing the flow rate by the area of the cylinder bore. The transients present in the input signal to the servo valve in a closed-loop system can disturb its steady-state operation. Typical performance specifications of servo valves are given in Appendix 1.

Steady-state Characteristics
Many parameters can be set to describe the steady-state behaviour of electro-hydraulic servo valves. Some of these parameters are the flow and the pressure conditions that exist in these servo valves. Flow gain and pressure gain are the two essential steady-state parameters for the servo valves. Other relevant parameters are the pressure drop across the valve, hysteresis, threshold (resolution),

linearity, and symmetry. The following sections describe these parameters and other associated terms of the servo valves.

Gain (G)
Gain is usually defined as the ratio of output to input. If the gain is high, its output varies rapidly to follow the command input, even in the presence of any external disturbance. Two types of gains are usually set for electro-hydraulic servo valves. They are (1) Flow gain and (2) Pressure gain. The flow gain and the pressure gain parameters for a servo valve can be determined from the flow curve and the pressure curve of the valve, respectively. The following sections explain the test setup to plot the flow curve and the pressure curve of a servo valve and the procedure to measure the flow gain and the pressure gain.

Flow Curve of an Electro-hydraulic Servo Valve
The schematic diagram of Figure 6.1 gives the setup to determine the flow characteristics of a 4-way closed-centre electro-hydraulic servo valve. The control flow in the valve is the flow passing through its control ports. In testing the 4/3-way servo valve, the flow moving out of one control port is assumed to be equal to the flow passing through the other port.

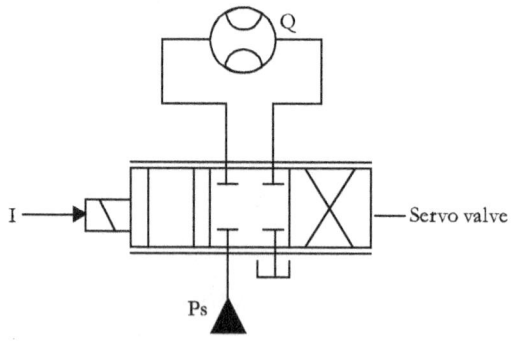

Figure 6.1 | A test setup to measure the flow gain

The flow curve of the electro-hydraulic servo valve is the representation of the flow rate (Q) through the valve versus the input current signal (I) to the valve, over its rated input current range. Figures 5.2(a), (b), and (c) show the typical flow curves for the under-lapped, zero-lapped, and over-lapped servo valves, respectively. As shown, the under-lapped servo valve allows the flow even when the spool is centred. In the over-lapped valve, the spool needs to be moved to a significant extent before the flow can start. The flow curve of a servo valve can be used to measure its flow gain, hysteresis, linearity, and symmetry.

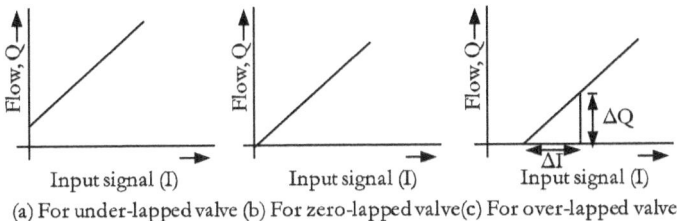

(a) For under-lapped valve (b) For zero-lapped valve(c) For over-lapped valve

Figure 6.2 | Flow curves of servo valves

Flow Gain of Servo Valve

The flow gain can be determined from its standard flow curve as given in Figure 6.2(c). It is given by the slope of the flow curve in any specified region and is the ratio of the change in the flow rate (ΔQ) to the change in the input current (ΔI), expressed in lpm/mA [(gpm)/mA]. The flow gain is given by:

$$\text{Flow gain, } G_f = \frac{\Delta Q}{\Delta I}$$

Internal Leakage of a Servo Valve

The total internal flow from the pressure port of a servo valve to the return port with its working ports blocked is the internal leakage of the valve. The internal leakage varies with the input current to the valve and reaches its maximum at the valve null (null leakage).

Pressure Drop across a Servo Valve

Figure 6.3 shows the setup for measuring the pressure drop and pressure gain of a 4/3-way closed-centre servo valve. Let the pressure at the port A of the valve be P_A and that at the port B of the valve be P_B. Pressure gauges installed at the ports A and B are used to measure the respective pressures at these ports. The load pressure drop (P_L) across the servo valve is the same as the differential pressure between the working ports A and B. It is given by:

$$P_L = P_A - P_B$$

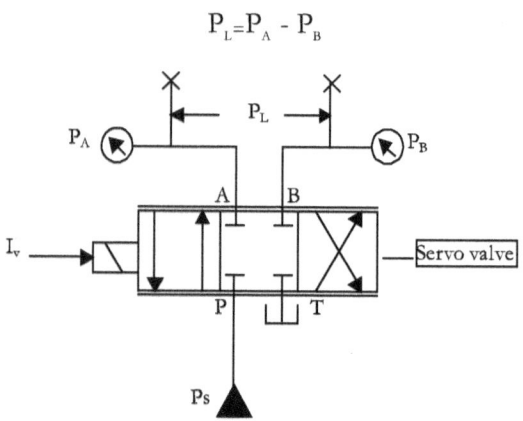

Figure 6.3 | A test setup to measure the pressure gain

Pressure Curve

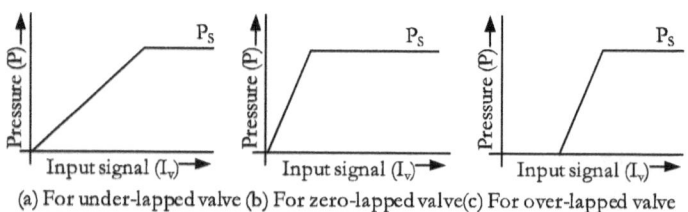

(a) For under-lapped valve (b) For zero-lapped valve(c) For over-lapped valve

Figure 6.4 | Pressure curves of servo valves

The pressure condition in a servo valve is dependent upon the lap condition of its spool. This fact is further explained with the help of the pressure curve, which is the graphical representation of the pressure (P) versus the input signal (I) to the servo valve. Figure 6.4 shows the typical pressure curves for the under-lapped, zero-lapped and over-lapped servo valves, respectively.

Pressure Gain
Pressure gain of a servo valve is defined as the change in its load pressure drop per unit input current in the absence of the control flow (with control ports blocked) through it. It is usually determined as the average slope of its pressure curve. That is,

$$\text{Pressure gain, } G_p = \frac{\text{Pressure drop across the load}}{\text{Input current}}$$

It is expressed in bar/mA [psi/mA]. The servo valve is to be pressurized at its null position with both the port A and the port B blocked. The pressure gain is measured by stroking the valve and recording the load pressure as a function of the input current.

Flow Vs Pressure Drop Curve
Figure 6.5 shows the typical no-load flow Vs the valve pressure drop characteristic of a servo valve.

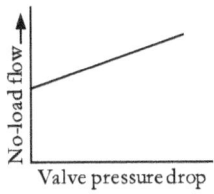

Figure 6.5 | A flow/pressure-drop curve of a servo valve

Hysteresis in Servo Valves
It is a measure of the inability of a servo valve to maintain the corresponding flow when the command input signal to the valve is increasing, and then it is decreasing. As shown in Figure 6.6, the

characteristic is plotted with an output, which is obtained by increasing the input current to its maximum, then decreasing the current, through zero, to its maximum in the negative direction, and then increasing back to zero. The hysteresis of the servo valve is specified as the maximum difference occurring over a complete cycle expressed as a percentage of its rated current.

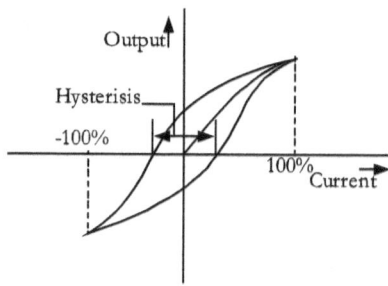

Figure 6.6 | The hysteresis of a servo valve

The hysteresis of the servo valve is an important performance parameter of the valve. Larger hysteresis can cause errors because it creates a null shift in the valve according to the direction from which the targeted output is approached. The hysteresis is very undesirable but inevitable in the servo valve. Typically, the hysteresis of the servo valve should be less than 3% of its rated current.

Threshold (or Resolution)
It is used to measure the smallest change in the command signal for which a servo valve can respond. In other words, it is the smallest change in the input current that must be applied to the servo valve to produce a measurable change in its output, expressed as a percentage of its rated current. The schematic diagram of Figure 6.7 shows the typical threshold parameter of a servo valve. It is usually measured by bringing the command signal applied to the valve to a particular level and then decreasing it until a change in the output flow is detected. A good-quality two-stage servo valve has a threshold of less than 0.5% of its rated current.

To improve the threshold of a flapper-nozzle valve, its spool is held in constant motion by superimposing a dither signal on the top of the command signal.

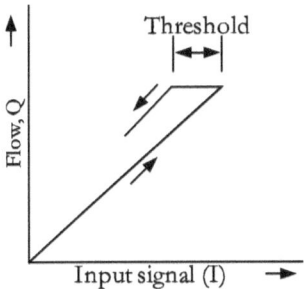

Figure 6.7 | The threshold parameter of a servo valve

Linearity and Symmetry

These are the other two parameters of the servo valve that are derived from its flow to input signal characteristic. They are: (1) Linearity and (2) Symmetry. The following bulleted lines define these characteristics:

- The linearity of the servo valve is a measure of the maximum deviation of the actual flow curve of the valve from that of its idealized flow curve. This relation is usually expressed as a percentage of its rated input current. Figure 6.8 shows this measure.

- The symmetry of the servo valve is a measure of the degree of equality between the average flow gain for one polarity of the input signal and that for its reversed polarity. Figure 6.8 shows this measure.

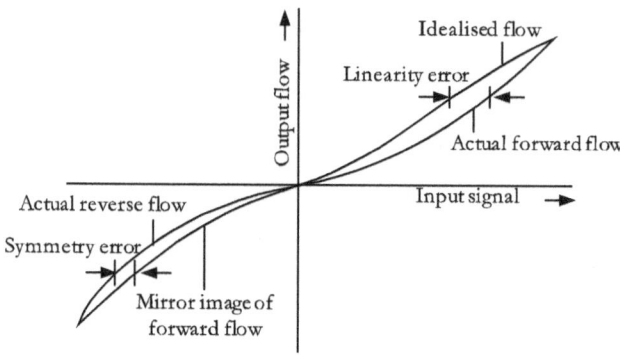

Figure 6.8 | The linearity and symmetry of a servo valve

Transient (Dynamic) Characteristics

All our discussion in the last few sections is related to the static characteristics of servo valves. The transient (dynamic) characteristics of the servo valve have a profound influence on the dynamic behaviour of the associated system.

If the transient response of the servo valve is slow, it may affect the performance of the system. An inherent resonance in the system can cause instability or oscillatory behaviour.

The dynamic response of the servo valve can be determined by analyzing their response to either a step input function or a sinusoidal input function.

The following section defines the step input function. The subsequent section explains the sinusoidal input response of the servo valve.

Step Input

Figure 6.9 provides the graphic representation of the unit step input. The unit step function is given by:

$$u(t) \quad = 1, t \geq 0$$
$$= 0, t < 0$$

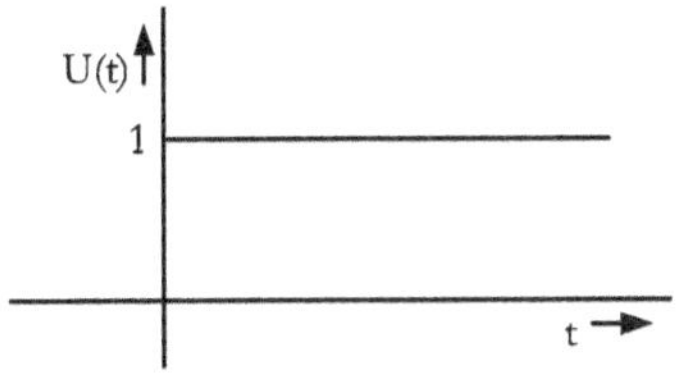

Figure 6.9 | A unit step input function

A step of height A can be written as Au(t).

Step Response

The transient behaviour of a servo valve can be represented by its step response, as demonstrated in Figure 6.10. It is measured by recording the spool displacement (x) versus time, for a specific step input current (I).

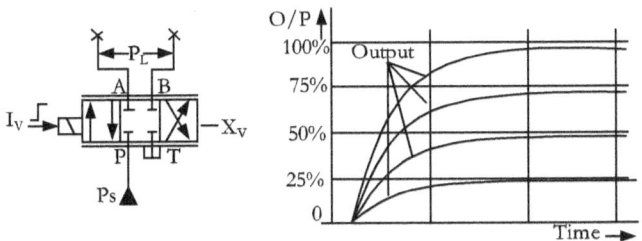

Figure 6.10 | A circuit arrangement for determining the transient response of an electro-hydraulic servo valve and typical responses.

Many servo valve manufacturers publish the step response characteristics of their valves. Usually, there are two possibilities when a step input is applied to the servo valve under no-load conditions. The system may appear as what is known as a first-order system or a second-order system. The following sections further explain these systems.

First-Order System

In the first-order system, the output of the valve increases much slower when a step input is applied to the servo valve but finally approaches the steady-state value. Figure 6.11 shows the step response of the first-order system.

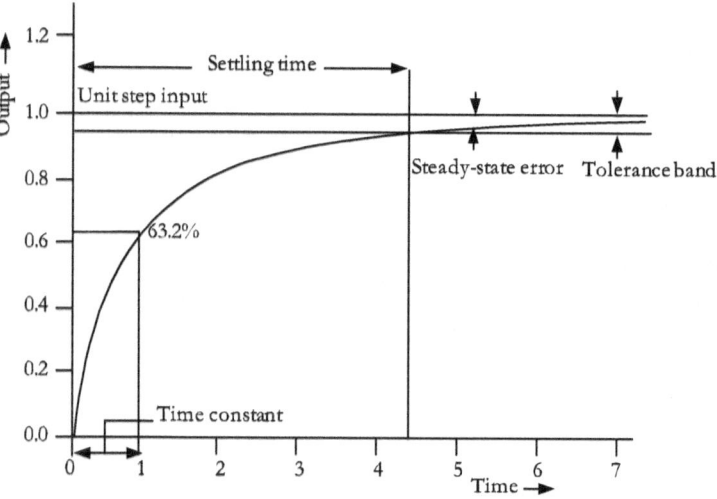

Figure 6.11 | A typical step response of a first-order electro-hydraulic servo valve system.

Two parameters are essential in the first-order servo valve systems. They are: (1) Time constant and (2) Settling time. The following bulleted lines define these terms:

- **The Time constant** of the first-order servo valve is the time required for its output to reach about 63% of its steady-state value. This value represents one measure of its time constant.

- **The settling time** of the first-order servo valve is the time required for its output to reach and stay within a defined tolerance band.

Second-Order System

The characteristic of a second-order system is shown in Figure 6.12. Its output increases rapidly, overshoot its steady-state value, and eventually settle at an appropriate value, in response to a step input function. Usually, an under-damped servo valve system exhibits this type of behaviour.

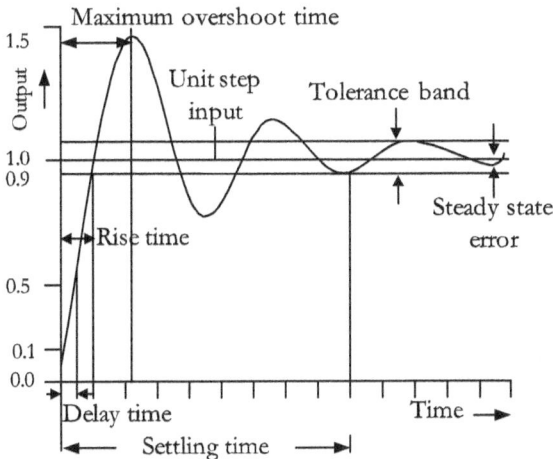

Figure 6.12 | A typical step response of a second-order servo valve system

The parameters related to the second-order servo valve system are: (1) delay time, (2) rise time, (3) maximum overshoot time, and (4) settling time. The following bulleted lines define these terms:

- **Delay Time:** It is the time required for the output of the servo valve system to reach 50% of its steady-state value.
- **Rise Time:** It is the time required for the output of the servo valve system to rise from 10% to 90% of its steady-state value.
- **Maximum Overshoot Time:** It is the time at which the maximum overshoot of the output of the servo valve system occurs.
- **Settling Time:** It is the time for the output of the servo valve to reach and stay within its stated tolerance band.

Sinusoidal Input (Frequency) Response
The step response of the servo valve is not considered as an exceptional measure of the performance of the valve, because of the inherent inaccuracies in measuring it.

A much more accepted method of measuring the transient response of the servo valve is by subjecting the valve to a sinusoidal input signal over a range of frequencies, and then comparing the behaviour of the output flow of the valve to the input signal, for each frequency.

When a sinusoidal input signal is applied to the valve, its output follows the input signal with some error. The amount of error depends on the frequency of the input signal. At low frequencies, the output follows the input signal. At high frequencies, the error tends to increase.

Figure 6.13 shows the typical input Vs output characteristics of the servo valve when subjected to a sinusoidal input.

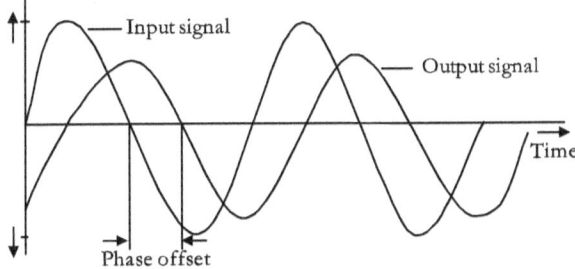

Figure 6.13 | The response of a servo valve to a sinusoidal input

The frequency response of the servo valve is usually measured with constant input current amplitude and a zero pressure drop across the load and is expressed by two parameters. They are: (1) Amplitude ratio and (2) Phase angle (Shift).

Amplitude Ratio (AR)
It is the ratio of the flow amplitude at any frequency to that at the specified reference frequency (typically 5 Hz). The amplitude ratio is usually expressed in decibels (dB).

That is, Amplitude ratio = $20 \log_{10}$ (AR) dB

Phase Angle (lag)
It is the difference in degrees between the phase of the sinusoidal input current and the corresponding phase of the output flow, measured at the specified frequency.

Servo valves are highly sophisticated devices that exhibit higher-order, non-linear responses. However, a first-order or a second-order expression is usually adequate, as it is only necessary to approximate the valve response to a relatively small frequency spectrum (50 Hz/100 Hz). Figure 6.14 shows the typical dynamic response of the second-order servo valve system with the Amplitude-Ratio curve and the phase-lag curve over some frequency range.

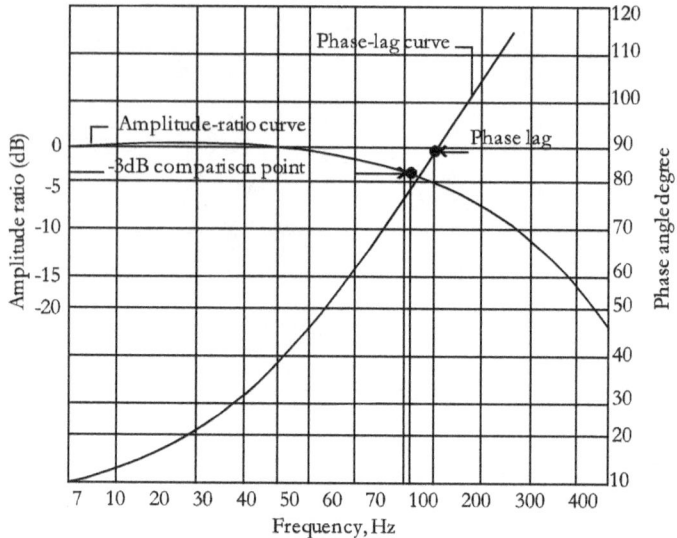

Figure 6.14 | A typical dynamic response of a second-order servo valve system

The dynamic response of the valve can easily be determined by referring to the frequency at which the −3 dB amplitude ratio (AR) and the 90° phase angle do occur. These values can be used to compare with the corresponding values of other servo valves of similar ratings.

Stability of Closed-loop Systems
The stability of a closed-loop system, together with adequate performance, is often difficult to achieve. Each component may perform correctly, yet the connection of the components into a closed-loop may result in hunting, oscillation, overshoot, chatter, sluggishness, poor resolution, or drift.

This type of load, length of hydraulic lines, sizing of valve and actuator, loop gains, presence of backlash, friction, load limiters, compliance, the location of feedback, or transducers can contribute to unacceptable closed-loop behaviour.

Chapter 7 | Typical Examples of Electro-hydraulic Servo Systems

As mentioned in an earlier section, the closed-loop electro-hydraulic servo valve systems can be classified as the position servo, the velocity servo, and the pressure/force servo. The following sections present some notable examples of these types of servo valve systems.

Electro-hydraulic Servo Positioning System

The primary objective of the position servo system is to move a hydraulic actuator in the system to the desired position and stop. This objective requires a closed-loop control system that has a command signal and a feedback signal. Once the actuator is in the desired position, the two signals produce a zero error signal (position error). Then, the servo valve must close to hold the actuator in place.

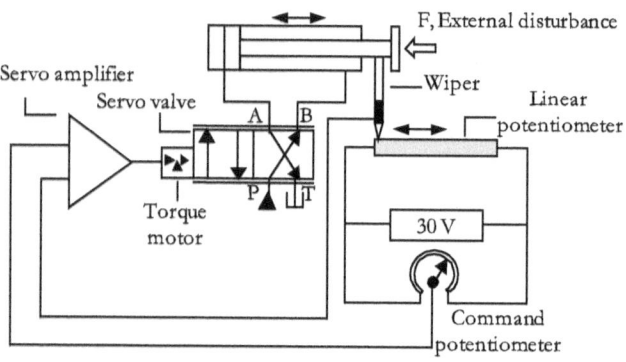

Figure 7.1 | A schematic diagram showing the arrangement of an electro-hydraulic position servo system

Figure 7.1 shows the schematic diagram of the electro-hydraulic servo system for positioning the hydraulic actuator. The system includes a servo valve, a servo amplifier, a command potentiometer, and a linear potentiometer. The desired position of the cylinder can be set by using the command potentiometer. The cylinder position is measured by the linear potentiometer, which gives an electrical voltage signal corresponding to the position of the cylinder, as the feedback signal. The servo amplifier compares the command signal with the feedback signal. The resultant error signal is amplified. It is then used to control the servo valve to adjust the position of the actuator to the set value. An electro-hydraulic position servo system with load position feedback can be used, where severe load resonance problem exists.

Electro-hydraulic Velocity Servo System

Figure 7.2 shows the electro-hydraulic closed-loop velocity servo system for maintaining the speed of a hydraulic motor to the set value.

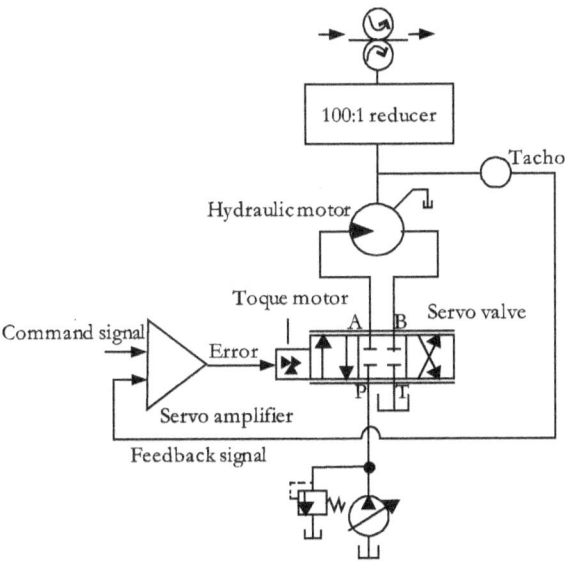

Figure 7.2 | A schematic diagram showing the arrangement of a typical electro-hydraulic velocity servo system

The system includes a servo valve, a servo amplifier, and a tacho generator. The desired speed of the motor can be set by using the command potentiometer. The tacho-generator is used as the feedback transducer for measuring the rotary speed of the motor and generating the feedback signal. The servo amplifier compares the command signal with the feedback signal. The resultant error signal is amplified. It is then used to control the servo valve. The servo valve must open enough to maintain the desired motor speed when the error signal (velocity error) becomes zero. Any deviation in the velocity of the motor from the set value generates an error signal that, in turn, causes the valve orifice to adjust itself to bring the velocity to the desired value.

Electro-hydraulic Pressure/Force Servo System

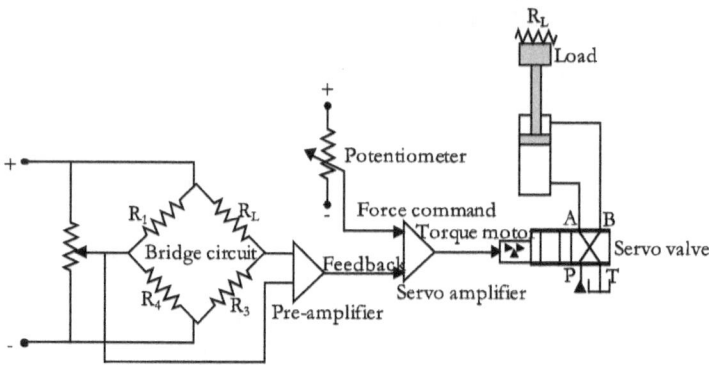

Figure 7.3 | A schematic diagram showing a typical electro-hydraulic force servo system

Figure 7.3 shows the typical electro-hydraulic pressure/force servo system for setting the pressure/force in the system to the desired value and holding it constant. It consists of a hydraulic cylinder, a servo valve, a pressure/force transducer, a command potentiometer, a servo amplifier, a pre-amplifier, and a bridge circuit. The pressure/force transducer R_L is mounted to sense the load pressure. Remember that force is a function of pressure for a given actuator. The bridge circuit is used for obtaining a feedback signal, which is pre-amplified before feeding to the servo amplifier. The cylinder in the force servo system increases the force on the load until the transducer output voltage equals the command voltage of the system. The system then carries this force until the command signal is changed. Any deviation in the desired pressure/force causes an error signal that will, in turn, cause the valve to open or close in such a way as to bring the pressure/force to the desired value.

Chapter 8 | Application of Servo Valves

The electro-hydraulic servo valves are used in high-precision systems employing high-power hydraulics controlled by low power (< 200 mW) electronics. The first use of servo hydraulics was in the aviation industry in the 1950s. Over time, industrial engineers applied servo-hydraulic techniques to their application. The use of the servo valves continued with military equipment, where they are used for radar drives, guidance platform drives, and missile launchers. The closed-loop electro-hydraulic servo systems are increasingly becoming the norm in machine automation, where the system requirements are demanding greater precision.

The servo valves also find applications in process plants, power generation, and mining.

The plastic manufacturing sector uses the power and precision of the servo valves for improving the quality of blow-moulded and injection-moulded parts.

The steel industry is a unique branch where the power of hydraulics is essential, along with the precision of electronics.

Typical applications also die casting machines, hydraulic press brakes, and animation and entertainment equipment.

The servo valves are also gaining full acceptance in a variety of application areas, such as material handling, oil exploration, and mobile equipment.

The servo valves provide an excellent static and dynamic response in large-tonnage high-performance hydraulic applications, especially in the areas of machine tools and primary flight controls.

Chapter 9 | Comparison of Proportional Valves and Servo Valves

After acquiring sufficient knowledge of the proportional valve and the servo valve systems, you are now in a position to compare these systems. Table 9.1 gives the comparison of the proportional valves and the servo valves.

Table 9.1 Comparison of the proportional and servo valves

Proportional valve	Servo valve
• The output is proportional to the input signal	• The output is proportional to the input signal
• The solenoid governs spool movement	• The torque motor governs the spool movement
• Operates in the open-loop or closed-loop modes	• Operates in the closed-loop mode
• The feedback on the spool position is obtained by using the LVDT	• The feedback on the output is obtained by using a potentiometer or a tacho-generator or a transducer
• Built to less stringent tolerances	• Built to the exacting tolerances
• Mass-produced, less expensive	• Custom-made, more expensive
• Spool overlap up to 20%	• Spool overlap up to 2%
• Moderate dynamic and static performances	• Excellent dynamic and static performances
• Less accurate than the servo valve	• Highly accurate than the proportional valve
• Less responsive	• Highly responsive
• Low maintenance requirements	• Higher maintenance requirements

10 | Objective Type Questions

1. Which type of hydraulic system is appropriate for the precise control of a high-speed, large-tonnage special application machine?
   a) Pure hydraulics
   b) Relay-based electro-hydraulics
   c) Proportional electro-hydraulics
   d) Electro-hydraulic servomechanism

2. To position its spool, an electro-hydraulic servo valve uses a(n):
   a) LVDT
   b) Encoder
   c) Torque motor
   d) Proportional solenoid

3. The distance the spool of an electro-hydraulic servo valve moves from its null position before the flow can start when a control signal is applied is called:
   a) Threshold
   b) Dither
   c) Offset
   d) Deadband

4. The electro-hydraulic servo valve system is a combination of:
   a) Solenoid valves, relays and feedback elements
   b) Solenoid valve, electronic controller, and transducer
   c) Torque motor, pilot spool, and main valve
   d) Torque motor, flapper nozzle, and main valve

5. Which of the following type of lap conditions in a servo valve has a deadband?
   a) Under-lap
   b) Over-lap
   c) Zero-lap
   d) Servo valve lap conditions do not exhibit deadband

6. Mark, the <u>incorrect</u> statement?
   a) The proportional solenoid is used in a servo valve to impart the necessary motion to the spool of the valve.
   b) The flapper-nozzle is an interface between the torque motor and the pilot spool in a servo valve.
   c) An under-lapped servo valve permits the flow even when the spool is centred.
   d) A torque motor provides a more linear force/current relationship than a proportional solenoid.

7. Mark, the <u>correct</u> statement
   a) An overlapped servo valve allows the flow even when the spool is centred.
   b) It is often necessary to readjust the null position of a servo valve whenever there is an output flow at zero input current.
   c) Hysteresis is a feature of servo valves to avoid errors.
   d) A torque motor in a servo valve can shift its main spool directly by using an LVDT.

8. The spool of a servo valve drifts from its centre position due to:
   a) The presence of contamination in the fluid
   b) The variations in temperature
   c) The variations in pressure
   d) All of the above

9. Mark, the <u>incorrect</u> statement about a servo valve
   a) The pressure gain is a steady-state characteristic.
   b) The maximum overshoot time is a second-order dynamic characteristic.
   c) The amplitude ratio is a frequency response parameter.
   d) The phase angle is a first-order dynamic characteristic.

11 | Review Questions

1. What is an electro-hydraulic servo valve?
2. Explain the operation of a closed-loop electro-hydraulic servo valve system, with a block diagram.
3. Classify the closed-loop electro-hydraulic servo valve systems.
4. Explain the operation of the electronic amplifier used in a servo valve.
5. Briefly explain the stages of power amplification in electro-hydraulic servo valves, with the help of block diagrams.
6. Describe the operation of the torque motor of an electro-hydraulic servo valve, with a simple sketch.
7. Why is a torque motor used in an electro-hydraulic servo valve rather than a solenoid?
8. Mention the methods used to connect the armature of a torque motor to the pilot spool of an electro-hydraulic servo valve for shifting its spool. Briefly explain.
9. Briefly describe the operation of the flapper nozzle arrangement in an electro-hydraulic servo valve, with a simple sketch.
10. Briefly describe the operation of the jet pipe arrangement in an electro-hydraulic servo valve, with a simple sketch.
11. Describe the operation of a typical two-stage electro-hydraulic servo valve. Explain with a simple sketch
12. What are the three types of lap conditions in electro-hydraulic servo valves?
13. Explain the lap condition in electro-hydraulic servo valves with a deadband.
14. Explain the line-to-line lap condition in an electro-hydraulic servo valve. What is its benefit over other types of lap conditions?
15. What is meant by the null drift in an electro-hydraulic servo valve? How is it adjusted?
16. Explain the method to measure the blocked line pressure gain of a 4/3-closed centre two-stage electro-hydraulic servo valve, with a test setup.

17. Explain the term 'static response' of an electro-hydraulic servo valve system.
18. Explain the term 'dynamic response' of an electro-hydraulic servo valve system.
19. Differentiate the static response and the transient response of an electro-hydraulic servo valve system
20. Explain the following terms in respect of servo valve characteristics: (1) Flow gain, (2) Pressure gain, (3) Hysteresis, (4) Threshold and (5) Step response.
21. What is the function of an electro-hydraulic position servo system?
22. Differentiate the position servo and the velocity servo electro-hydraulic systems.
23. Describe the operation of a velocity servo system, with an appropriate example.
24. Describe the operation of a position servo system, with a proper example.
25. Compare the electro-hydraulic proportional valves and the servo valves.
26. Mention any three applications of electro-hydraulic servo valves in the industry.

Objective type questions - answer key:
1-d, 2-c, 3-c, 4-d, 5-b, 6-a, 7-b, 8-d, 9-d

Appendix 1

Typical Performance Specifications

Table A1.1 | Performance specifications of servo valves

Parameter	Typical value
Maximum operating pressure, ports P, A, B	350 bar (5000 psi)
Nominal flow rate, Q @70 bar (1000 psi)	3.8, 5, 9, 10, 19, 20, 28, 38, 40, 45, 47, 57, 76 lpm [1, 2.5, 5, 7.5, 10, 12, 12.5, 15, 20 gpm]
Null leakage flow @P=140 bar (2000 psi)	<3% of Q
Hysteresis	<0.2%
Threshold	<0.1%
Thermal drift ($\Delta T = 50°C$)	<1.5%
Response time	≤ 12 ms
Vibration on three axes	30 g
Ambient temperature range	-20 to +60°C
Fluid temperature range	-20 to +80°C
Fluid viscosity range	5 to 400 cSt
Recommended viscosity	25 cSt
Fluid cleanness (ISO 4406)	17/15/12
Integrated electronics and position feedback	Yes/No
Rated size	ISO 4401-03 (CETOP 03)
Spool with zero overlap	
Seals	NBR / FPM
Reference signal	±10V
Rated current, maximum	200 mA
Valve mounting pattern	NFPA (D05)

Appendix 2

PID Control

PID controllers are widely used in closed-loop (feedback) control of industrial processes. A PID controller consists of elements with the following three functions. They are: (1) Proportional (P), (2) Integral (I) and (3) Derivative (D).

At first, a designer of a process/system must know its characteristics to control pressure, flow, level, temperature, etc. He must then decide the type of controller to solve the control task. A PID controller is the primary tool in most closed-loop control systems.

A Simple Closed-loop Control System

Figure A2.1 shows the block diagram of a simple closed-loop system to control a machine. The control system consists of a proportional gain amplifier and a summing point. The desired operating point that is to be maintained is called the set point (SP). The operating information from the system is called the process variable (PV) or measured variable.

The process value and the set value is compared (by subtraction) to produce a difference called the error. The error is amplified by a proportional gain factor Kp. The amplifier then attempts to adjust the output to reduce the error. In short, the closed-loop system measures, compare and then adjusts.

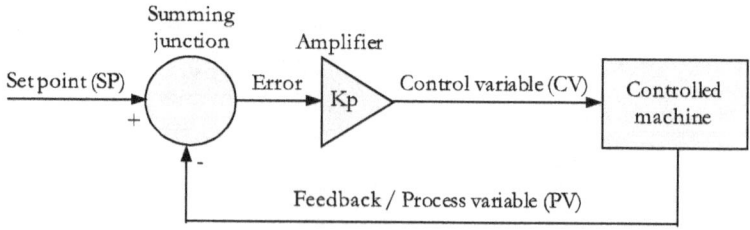

Figure A2.1 | Simple closed-loop system

A simple closed-loop system with a proportional function will rarely work correctly. The proportional gain Kp must be very high to bring the machine's operating point near to the set point. However, when a high gain is used, the system tends to be unstable. As the system tends to approach its set point, the error value also decreases. That means, the system tends to approach the set point at a slower rate. At a steady state, the output will level off at some value that is less than the set point and will never reach the desired output. The difference between the steady-state operating point and the desired operating point is called offset. (Figure A2.2)

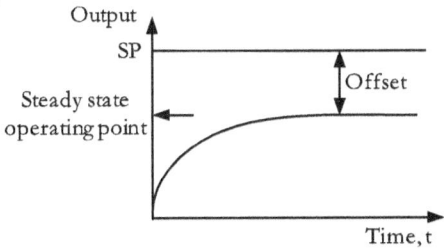

Figure A2.2 | Output response with low proportional gain

It is necessary to add two more functions to the closed-loop system to achieve more desirable levels of performance. They are: (1) Integral function and (2) Derivative function. Each of the functions, namely the proportional, integral, and derivative functions, serves a specific and unique purpose in the system.

Integral Function
The integral function is denoted as $K_i\int$, where K_i is a multiplying constant called the integral gain constant. The integral function integrates (sums up) the input (offset/error) over time, and then the result is multiplied by the parameter K_i.

For a small offset, the integral function will accumulate slowly, and its output will increase only gradually. For a large offset, the output of the integral function will change more rapidly to reduce the offset.

In a closed-loop PID control system, the offset can be reduced to near zero by increasing the integral gain constant K_i to some positive value. However, a system with an excessive amount of Ki will overshoot and oscillate.

Derivative Function

The derivative function is denoted by $K_d \frac{d}{dt}$, where K_d is a multiplying factor. The derivative function differentiates the input, and the result is then multiplied by the parameter K_d. For example, when a linear ramp (constant slope) is input to the derivative function, it will output a voltage that is equal to the slope of the ramp. Further, the derivative function will output zero, when a constant DC voltage (slope = 0) is input to it. For more complex waveforms, an approximate derivative can be found by sampling.

The derivative function tends to dampen the acceleration/ deceleration rates. Consequently, the tendency of the system to oscillate is reduced, and the system tends to settle more quickly. In other words, by increasing the derivative constant K_d, the transient response of the system can be improved by eliminating the overshoot and hunting. However, remember that the derivative constant does not affect the offset.

A Typical PID Configuration

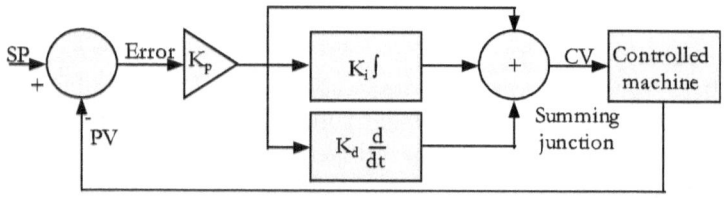

Figure A2.3 | A typical PID configuration

The most commonly used PID configuration for a closed-loop control system is shown in Figure A2.3. The output of the controller can be given as:

$$\text{Output (CV)} = Kp\, e + Kp\, Ki \int e\, dt + Kp\, Kd\, \frac{de}{dt}$$

Even though it is possible to get a faster response by further increasing the proportional and derivative gain parameters K_p and K_d, potentially damaging results may occur due to the development of excessive voltage, current, and force transients.

The Behaviour of a PID Controller

Typical behaviors of a PID controller for various values of parameters Ki, Ki, and Kd are given in Figure A2.4:

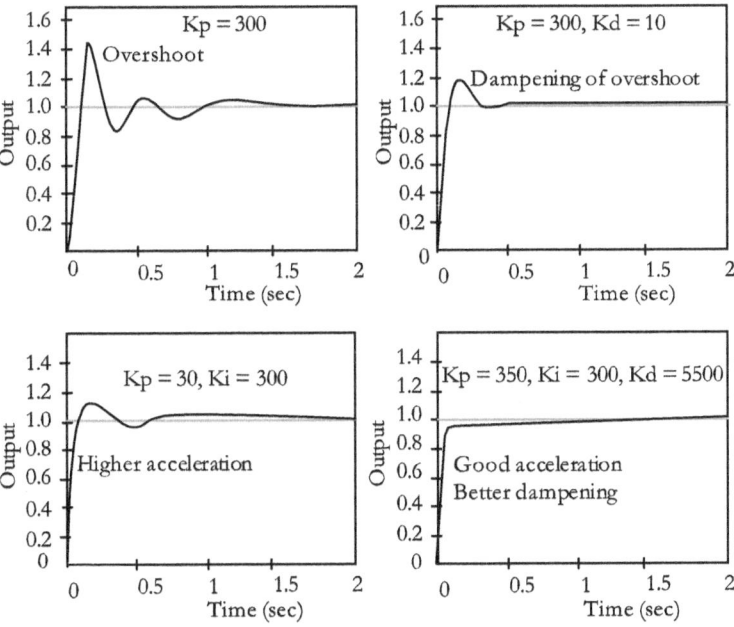

Figure A2.4 | The behaviour of a PID controller

Tuning of PIDs

Tuning of PID controllers involves selecting the best values for K_p, K_i, and K_d. to achieve results that are closer to the desired performance.

The essential characteristics of PID adjustments are as follows:

- Increasing the proportional gain constant (Kp) will result in a faster response and will reduce the offset. However, increasing the proportional gain will cause overshoot and hunting.

- Increasing the derivative gain constant (Kd) will reduce overshoot and hunting. However, it will reduce the offset.

- Increasing the integral gain constant (Ki) will cause the PID to reduce the offset to near zero.

For tuning a PID, the designer must be very familiar with the characteristics of the system. The system must be accurately modelled as far as possible by determining the mechanical and electrical parameters of the system.

Further, the desired response of the system to changes in the set point or loading conditions must be evaluated. The designer must also be aware of the theoretical aspects of the PID functions.

12 | References

1. Article on: 'A BRIEF HISTORY OF ELECTROHYDRAULIC SERVOMECHANISMS', R. H. MASKREY and W. J. THAYER, MOOG Industrial Controls Division, Moog Inc., NY, USA
2. Article on: 'TRANSFER FUNCTIONS FOR MOOG SERVOVALVES', W. J. THAYER, MOOG Industrial Controls Division, Moog Inc., NY, USA
3. Bulletin HY14-1483-M3/US 'Installation Guide, Model 23-7030, Servo valve Control Amplifier', Parker Hannifin Corporation, Ohio, USA
4. Document on: 'Electro-hydraulic Valves... A Technical Look', MOOG Industrial Controls Division, Moog Inc., NY, USA
5. Document on: 'THEORY AND APPLICATION OF SERVO VALVES', DYNAMIC TESTING EQUIPMENT, Inc
6. Document on: 'VSC4 4-WAY SERVO VALVES', Oilgear, Milwaukee, WI, USA
7. John R. Hackworth and Frederick D. Hackworth, Jr., 'Programmable Logic Controllers Programming Methods and Applications' Pearson Education
8. Materials on: 'MOOG G761 Series Installation and Operation Instruction Electro-hydraulic Servo valve', 'TECHNICAL BULLETIN 117, SPECIFICATION STANDARDS FOR ELECTROHYDRAULIC FLOW CONTROL SERVOVALVES', MOOG Industrial Controls, Moog Inc., NY, USA

Fluid Power Educational Series Books

1. Pneumatic Systems and Circuits -Basic Level (In the SI Units)
2. Industrial Pneumatics -Basic Level (In the English Units)
3. Pneumatic Systems and Circuits -Advanced Level
4. Electro-Pneumatics and Automation
5. Design of Pneumatic Systems (In the SI Units)
6. Design Concepts in Pneumatic Systems (In the English Units)
7. Maintenance, Troubleshooting, and Safety in Pneumatic Systems
8. Industrial Hydraulic Systems and Circuits -Basic Level (In the SI Units)
9. Industrial Hydraulics -Basic Level (In the English Units)
10. Hydraulic Fluids
11. Hydraulic Filters: Construction, Installation Locations, and Specifications
12. Hydraulic Power Packs (In the SI Units)
13. Power Packs in Hydraulic Systems (In the English Units)
14. Hydraulic Cylinders (In the SI Units)
15. Hydraulic Linear Actuators (In the English Units)
16. Hydraulic Motors (In the SI Units)
17. Hydraulic Rotary Actuators (In the English Units)
18. Hydraulic Accumulators and Circuits (In the SI Units)
19. Accumulators in Hydraulic Systems (In the English Units)
20. Hydraulic Pipes, Tubes, and Hoses (In the SI Units)
21. Pipes, Tubes, and Hoses in Hydraulic Systems (In the English Units)
22. Design of Industrial Hydraulic Systems (In the SI Units)
23. Design Concepts in Industrial Hydraulic Systems (In the English Units)
24. Maintenance, Troubleshooting, and Safety in Hydraulic Systems
25. Hydrostatic Transmissions (HSTs) (In the SI Units)
26. Concepts of Hydrostatic Transmissions (In the English Units)
27. Load Sensing Hydraulic Systems (In the SI Units)
28. Concepts of Load Sensing Hydraulic Systems (In the English Units)
29. Electro-hydraulic Proportional Valves
30. Electro-hydraulic Servo Valves
31. Cartridge Valves
32. Electro-hydraulic Systems and Relay Circuits
33. Practical Book: Pneumatics - Basic Level
34. Practical Book: Electro-pneumatics - Basic Level

35. Practical Book: Industrial Hydraulics – Basic Level
36. Programmable Logic Controllers and Programming Concepts
37. Compressed Air Dryers

For more details, please visit: **https://jojibooks.com**

About the Author

Joji Parambath is a trainer in the field of Pneumatics, Hydraulics, and PLC, for over 25 years. During his career, he has trained numerous professionals from the industries as well as faculty members and students of engineering institutions.

At present, he is the key trainer at Fluidsys Training Centre, Bangalore, India, (https://fluidsys.org), which provides training in Pneumatics and Hydraulics. He has already written two books on Pneumatics and Hydraulics. The publication of the present series of 32 books is intended to restructure and update the existing books.

The author wishes to thank all trainees for their lively interaction and many useful suggestions during the training programmes that prompted the author to write the present series of books. You may send your feedback to joji.p@hotmail.com

10th June 2020